Practice
Photography for Surveyors

Gareth W Evans

Routledge
Taylor & Francis Group

LONDON AND NEW YORK

First published 2005 by Estates Gazette

Published 2014 by Routledge
2 Park Square, Milton Park, Abingdon, Oxon OX14 4RN
711 Third Avenue, New York, NY 10017, USA

Routledge is an imprint of Taylor & Francis Group, an informa business

ISBN 13: 978-0-7282-0477-5 (pbk)

Commissioning Editor: Alison Richards
Typesetting: Isabel Micallef — in Meta and Palatino
Cover and interior design: Deborah Ford, EG
Cover image: Gareth Evans

Contents

Acknowledgements

Thanks are due to my wife, who translated my overtechnical outpourings into readable English.

Preface

After taking tens of thousands of photographs of buildings over the years, I realised that I knew rather a lot about it. I looked at photographs by other surveyors and realised, too, that many of them could do with some help. So here it is.

Introduction

There are many books about photography. Most of them will tell you, in some detail, how to photograph flowers, pets and people, not to mention rapidly moving vehicles. They deal mainly with the creation of artistically satisfying images and will also give plenty of detail of the mechanical and electronic processes involved. This book is *solely* about photographing buildings. It is aimed at the busy surveyor, or other property professional, who wants fuss-free but high-quality pictures for reports, filing, e-mailing or web-posting. Technical detail is kept to the minimum necessary.

A glance at any property magazine, such as *Estates Gazette* or a textbook such as *Structural Surveys of Dwelling Houses* (Ian A Melville, Ian A Gordon, Paul G Murrells, Estates Gazette, 1992), shows the importance of photographic images to the surveying profession. Many of these images have been taken by professionals. However, there is absolutely no reason why the surveyor, with only reasonable equipment, cannot produce excellent images that are capable of great enlargement and which will enhance a report (figure 1), provide an *aide-mémoire*, or can be attached to a database.

This small publication attempts to provide the minimum knowledge that a property professional should have in order to produce quality images. I have concentrated on the practical requirements of the onsite property professional. Cameras are now very sophisticated. The vast amount of theory that used to be a necessary prerequisite for the production of quality images is now less necessary for most purposes. This is not a book for a photo enthusiast who wishes to carry around a whole bag of lenses, tripods and filters.

Figure 1: Subsidence distortions with (inset) close-up. Such images can enhance your reports

Photographic images are thrust upon us daily. Most of the time, they are taken by professionals and we take them for granted. We have all seen the type of amateur photograph that is out of square, out of focus, blurred, too dark, washed out, or even cuts off the subject's head. In many cases, these images are poor because the hopeful photographer is trying to exceed the capabilities of the equipment. In other cases, the images *are poor because the photographer is trying to exceed his or her own capabilities.*

There is no reason why, even with a relatively basic camera, one should produce such awful pictures. Working within the limits of the equipment, and with a little knowledge, one can produce images, which, if not artistic, are at least sharp, correctly exposed, and show the subject properly.

I have deliberately avoided the recommendation of any particular model of camera since the pace of change, especially in digital photography, is so great that all current models will be obsolescent or out of production by the time this publication reaches print. However, I have used images of particular cameras to illustrate a type of camera.

I have no formal photographic training. However, many years ago, I learned not only how to take pictures but also to develop and print black and white photographs and to process colour slide film. Since then, I have used many cameras and, a few years ago, abandoned the use of film for most purposes in favour of digital photography. A quick glance at the folders on my hard disk shows that, in 2004, I took approximately 16,000 digital images for business purposes. A friend jokes that I am the only person he knows who actually wears cameras out! I hope to pass on the benefit of all this practice.

Finally, there are three basic essentials:
● When you acquire a new camera, read the camera instructions right through.
● Read them again the next day.
● Practice.

At the time of writing, the big photographic dichotomy is between film and digital processes, with digital cameras now outselling film cameras despite the higher initial cost. This section introduces the fundamentals of imaging and underlines the advantages of digital photography.

The photographic process

Film

In order to appreciate the benefits of digital cameras for property work, some understanding of film cameras is essential. In film cameras, the image is projected onto a thin, clear, strip of film that bears a photosensitive coating, which is based on silver salts. The action of light changes the chemical properties of the coating to what is called a "latent image". This cannot be seen, but must be "developed" by a chemical process to produce a visible image. This is a negative image, which, in crude terms, shows white as black, black as white, and colours as different colours. In addition, the presence of an overall orange masking tint makes it less easy to work out the original colours from the colour negative.

Conventional film produces one negative of the image. With care, an indefinite number of prints can be made from this negative but it is a delicate object and must be carefully protected. Any prints, or copies of subsequent prints, inevitably lose some of the original information. Images must be stored physically in a logical order. In the context of surveying, prints must be stuck into documents, and if there are many copies of the document, there must be many copies of the print made by the photo processor. Alternatively, a transparency, a positive image, can be produced. This suffers from the same limitations of copying, and moreover requires inconvenient viewing arrangements.

Most film is 35mm, ie it is a strip 35mm wide, which is a standard produced in the 1920s by Leica, originally for testing movie film. Other formats come and go; professionals often use "6 x 6" film (producing images 60mm square) or larger.

Individual film images are commonly termed "frames" (the frames are numbered on the film margin for reference). As a former 35mm user, I use this term also for digital images, but others may call them "images".

Film products require careful storage if they are to be retained for the long term. Both heat and x-rays degrade the film, as does time. In addition, different types of film are needed for daylight and interior light (see page 15).

Digital

In digital cameras, the operation of the lens is identical in principle to that in a film camera. However, the image is projected onto a photosensitive array, which will be either a CCD (charge-coupled device), or a CMOS (complementary metal-oxide-semiconductor) of many individual elements, each of which attributes colour and brightness to the small portion of the image that falls upon it.

The resulting overall image, composed of several million elements (pixels), is compressed within the camera's microprocessor and stored electronically, usually in "flash" memory. It can be copied, in theory, an infinite number of times. At each copying, barring defects in equipment, there is no loss of information. The image can be readily transmitted electronically by e-mail, archived to electronic or opto-electronic media, and indexed by some form of database. It can be embedded seamlessly into documents and as many copies as required can be printed in your own office. The image can be retrieved to your computer screen rapidly, or "burned" to CD or DVD and viewed on a television using the DVD player. In addition, digital images are ideal for computer presentations.

Provided the number of megapixels is adequate, digital cameras now match or exceed the image quality of an equivalent film camera.

Most cameras now comply with the EXIF standard (Exchangeable Image File Format), which stores interchange information in image files, especially those using JPEG (see page 9) compression. The most useful feature of this may be the date (provided that you have set up the camera correctly).

For office use, the advantage of the digital camera is clear. The difference in cost is greatly outweighed by the absence of expenditure on film and processing. Consequently, this book deals with digital camera work only.

In this chapter, the basics of camera choice are explained.

Camera types

Traditionally, film cameras fall into two groups: compact and SLR. Compacts are self-contained, with a single lens. SLR (single-lens reflex) cameras (figures 2 and 3) have interchangeable lenses and the image is viewed, by a mirror / prism system, through the lens up to the moment of exposure.

The digital equivalent of the 35mm compact will usually have a 3:1 zoom lens or better, and a direct vision viewfinder (figures 4 and 5).

Digital cameras differ from film cameras in that there is a rather narrow middle ground between the

Figure 2:
Professional
digital SLR

Figure 3: Consumer digital SLR

Choosing the equipment

Figure 4: Basic compact

Figure 5: High-specification compact

Figure 6: Prosumer compact with long zoom and through-lens viewfinder

Figure 7: Typical SLR zoom lens

equivalent of the 35mm compact and the SLR. In this sector of the market, there are several highly specified models (for instance, figure 6) with fixed, but wide-ranging, zoom lenses and a (video) viewfinder that looks through the lens.

The SLR has interchangeable lenses of very high quality (figure 7) and probably a higher-quality CCD sensor (6 megapixels — MP — at least). It is now possible to buy an SLR and short-zoom lens for under £1000.

Virtually all models have an LCD (liquid crystal display), which permits an instant check of the image

that has just been taken, or which can be used as a viewfinder. While, except for the most expensive models, the ultimate quality of the image does not approach that of the equivalent film SLR and correctly processed film, the advantages of digital images may be considered more significant in terms of insertion into computer-generated reports, databases and archives. It is also possible to take as many pictures as is necessary to ensure adequate coverage of the building, without concern for cost. Perhaps best of all, there is no need to worry about image adequacy, as instant checking is possible.

Selecting the camera

The personal touch

If you expect to take a lot of pictures, then it is important that you are comfortable with your camera. In some cases, this may mean that a camera that "feels" right in your hand should be preferred to one that offers more facilities.

Do not buy a camera by mail order, unless you have already tried it. The result could be very disappointing. You may find that the focusing or zooming buttons or rings do not fall comfortably to your fingers; that the camera does not fit satisfactorily against your face as you use the viewfinder, and so on. Some people prefer a lightweight camera, while others prefer a reassuring chunkiness in the hand.

Battery type

Depending on your operational requirements, battery type may influence your choice. Lithium-ion batteries are commonplace but are expensive if lost or damaged. However, if you need to use the camera heavily, they can usually be recharged quickly (about an hour), so that a spare is not necessary unless you are shooting continuously or are away from an office or hotel room.

Cameras that use AA batteries drain them rapidly, but if you run out in any urban location, standard AA replacements can easily be picked up.

What is your objective?

If you intend to take basic record images, which serve only as a bare illustration of the building or other feature, you should not need to spend too much unless you expect to take many of the pictures in poor light.

If any part of the image is critical to the understanding of the report or other document, or if it is likely that the images may be referred to for reference purposes in the future, then you need to invest in a superior camera, possibly an SLR.

Camera images supplied by Canon UK.

To get the best from your camera, a little technical information is necessary. This section sets outs the essential knowledge that all serious photographers need.

How cameras work

Critical elements of a camera

The most critical element of any camera is probably the lens. Computerised lens production has resulted in the production of high-quality lenses at remarkably low real cost. However, examination of even high-quality images will reveal minor departures from perfection. Light falling upon the camera lens from an image is diverted, or refracted, within the lens and at its boundaries. In simplistic terms, the light rays are focused through a notional point within the lens and form a small, inverted image on the receiving film or CCD array behind. Imperfections in the lens design mean that this recorded image is slightly degraded from the original. This is the first point in the photographic system at which the information in the real-world object is degraded. Any subsequent processing of the image may then degrade the information further.

The other critical element is the recording medium of film or CCD. This must be carefully aligned at right angles to the optical axis of the lens (an imaginary line about which the lens is symmetrical) so that the image projected by the lens falls most accurately on it. The projected image is, in reality, not precisely flat due to the limitations of lens technology but, in practice, this is not noticeable.

These two elements must be encased in a light-tight box. There must be a contrivance that allows light through the lens at the appropriate time, and a system for recording images successively. Virtually all cameras also include a device for ensuring that the amount of light that falls through the lens is correct.

Beyond these elements, film and digital cameras begin to depart significantly from one another. However, at present they look very similar externally.

Exposure and the principle of reciprocity

More about the lens

A camera lens has two main characteristics: its focal length and its aperture.

The "focal length" of a lens is the distance between the centre (or, strictly, optical centre) of a lens focused on infinity and its focus point. If the focus point is on the CCD, the lens is "in focus".

Camera lenses are not like a simple magnifying glass lens, but are made of more than one piece (or element) of glass, typically five or more (to minimise distortions). The focal length is quoted as an effective equivalent from the optical centre.

The "aperture" is the size or diameter of the lens; again, with complex lenses an equivalent figure is given. The aperture is represented as a ratio of the focal length of the lens (conventionally, referred to as f), such as $f/4$.

Figure 8: Colour fringing is usually apparent against a dull sky

Figure 9: The full image shows how, in practice, colour fringing is not usually a nuisance

relationship. The higher figures (ie the smaller apertures) are produced by "stopping down" the lens by means of an iris of overlapping metal plates. For this reason, older photographers will speak of an "f stop". Lenses of wide aperture are termed "fast".

Most lenses produce the sharpest result in the mid-section of their aperture range, and you should bear this in mind for critical work.

Zoom lenses (discussed below) are now fitted to most compact cameras and are commonly provided as an option for cameras that take interchangeable lenses. They feature a range of focal lengths and are motorised or operated by a sliding or rotating collar. The optical quality is less than for a lens of fixed focal length. This is because of the design compromises that have to be made in order to achieve a workable size and weight for the zoom lens. However, modern design is so good that this lower quality is hardly detectable. Also, zoom lenses tend to have slightly smaller apertures than fixed lenses.

Colour fringing and distortion

No lens is perfect. If you examine your images carefully (especially off-centre), you may find that objects (particularly those with a pale background) have a purplish fringe on one side and a yellowish fringe on the other (figures 8 and 9). This is due to the inability of the lens to refract different frequencies of light equally, despite the most cunning computerised design.

Images, especially those taken at wide-angle settings, may show distortion of straight lines towards the edges of the frame. If the lines bulge away from the centre of the image, that is "barrel distortion"; if they curve inwards, that is "pincushion distortion" (figures 10 and 11). Some lenses are more prone to these problems than others, depending on the compromises made by the designers.

The shutter

In modern digital cameras, the shutter is a mechanical (usually now electronically controlled) or electronic arrangement for ensuring that the light from the scene in front of the lens reaches the recording medium for the desired length of time. Some cameras have both mechanical and electronic arrangements.

Shutter sounds

Digital cameras are quiet in operation, although some may have noisy zooming motors. Most cameras produce a range of bleeps or other sounds to show that various things are happening (even a simulated shutter sound!). If you wish to take pictures, for instance, of tenants in breach of lease covenants, then it may be advisable to turn off all sounds on the camera, if that option exists.

This means that the aperture is one-quarter of the focal length. The smaller the number after the f, the larger the aperture and its light-gathering ability. Modern camera lenses have a maximum aperture typically from f/2 to f/5.6, depending on the type of camera. Nowadays, the number is commonly written as f2, and so on.

The f numbers follow a regular series, conventionally f/1.4, f/2, f/2.8, f/4, f/5.6, f/8, f/11, f/16, f/22 and f/32 (although current cameras often include only a limited range, say, from f/2.8 to f/11). These figures are rounded, but there is an obvious mathematical

Figure 10: Pincushion distortion (simulated)

Figure 11: Barrel distortion (simulated)

Exposure

To create an optimally exposed image, a certain amount of light has to fall on the CCD. Broadly speaking, it does not matter whether this is a lot of light all at once, or less light over a long period. Therefore, a correct exposure could be, for instance, 1/250 second at f/4, or 1/500 second at f/2.8. It may help to visualise this in terms of a short, fat cylinder of light or a long, thin cylinder of light (figure 12). This relationship is termed "reciprocity" (see below).

There is, conventionally, a recognised range of shutter speeds, which are given as rounded figures, each of which is half or double the next. Conventional mechanical shutters have a range of fixed speeds (which, in reality, may deviate significantly from the nominal figure). The typical range is 1/8 second,

1/15, 1/30, 1/60, 1/125, 1/250, 1/500, 1/1000; better cameras will extend the range in both directions. In practice, modern cameras may automatically select shutter speeds, such as 1/438 second. Fast shutter speeds are better at freezing movement, while the choice of aperture has other significance, which is discussed below. A "fast" lens allows the use of higher shutter speeds and the taking of pictures in poorer light.

The principle of reciprocity is utilised by the metering systems in virtually all modern cameras, so that if the photographer chooses a fast shutter speed the camera can set a corresponding wide aperture. Similarly, the photographer can choose a wide aperture and the camera can set a correspondingly high speed. For instance, 1/60 second at f/8 is the same as 1/125 at f/5.6 in terms of light reaching the CCD.

More sophisticated cameras will offer a range of options, including full automatic operation ("point-and-shoot"); programmed exposure that can be left to operate automatically or can be varied; shutter priority (the shutter speed is deemed more critical); and aperture priority (the aperture is more critical). Outside the range in which reciprocity operates, the system may not be able to expose the CCD satisfactorily, giving rise to muddy or noisy images. This is termed "reciprocity failure". In virtually all modern cameras (excluding the simple disposable "party cameras"), all this calculation is dealt with automatically by a built-in metering system, which measures the light reflected from the subject and calculates the exposure. The system automatically deals with the reciprocity issue, within the limits of the camera's capability.

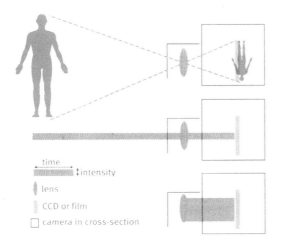

Figure 12: Exposure reciprocity

Figure 13: Adelaide House and
London Bridge on wide-angle setting

Despite the powerful and flexible tools included in most digital imaging applications, there is still no substitute for correct initial exposure. Perfect exposure should produce an even balance of detail in the highlights and the shadows. Too much light will burn out the highlights and lose the details, and too little light will result in total loss of detail in shadows. More sophisticated digital cameras can produce a histogram, which will immediately show you whether the exposure is satisfactory.

The focal length of the lens

Some of the terminology used to describe lenses can be confusing. To understand this, an aside on film photography is needed. A 35mm SLR is normally sold with a so-called "standard" lens of 50mm focal length. This is usually described as giving an angle of view roughly equivalent to that of the human eye. Lenses of shorter focal length are termed "wide-angle", while those of longer length are termed "long-focus", or more commonly "telephoto". Cameras with larger film size, such as 6 x 6cm, have correspondingly longer equivalent lengths. Cheaper compact cameras typically have a "semi-wide" of 35mm or 38mm as a fixed lens.

In practice, all except the cheapest cameras (both film and digital) come with "zoom" lenses, with a variable focal length, 3:1 being the most common range. The major strength of the SLR is its capacity to take a whole range of different lenses, which fit into an accurately machined bayonet seating on the lens ring at the front of the camera. Such lenses can range from an extreme "fisheye", which shows a hemispherical field, to ultra-long (and very heavy) telephotos, favoured by sports and nature photographers. SLRs are often sold as a

package with a zoom, but usually you will have to pay extra. Zooms tend to be short (for instance, 28–90mm range) or long (typically, 75–300mm).

Digital cameras have a smaller receiving area (the CCD) for the incoming light than a film camera and the focal lengths are correspondingly shorter. The traditional focal lengths are so entrenched in the photographic industry that the lenses are frequently quoted as their 35mm equivalents. Therefore, an 8–24mm digital camera zoom equates to a 35–115mm zoom for a 35mm camera. In this publication, I use 35mm lengths, unless otherwise stated.

Shorter focal lengths:
- Allow a wider view angle.
- Have a greater depth of field.

Longer focal lengths:
- Have a narrow angle of view.
- Have a limited depth of field.

Careful choice of aperture and focal length combination will allow you to control, within the limits, both the area of the subject on the image and the depth of field. A rectangular room can usually be fully recorded on two frames (views from opposing ends of the room) with a 28mm lens, while to obtain acceptable frame-filling images of the upper details of a tower block will require a lens of 200mm focal length or longer (figures 13 and 14).

Depth of field

As the lens aperture is stopped down, the angle at which the light rays fall on the recording medium changes. This increases the distance in front of the camera at which subjects appear to be acceptably sharp in the final image. This phenomenon can be used to ensure that your property subject appears

Figure 14: Adelaide House on telephoto setting

sharp overall. Alternatively, you may wish to blur the background or foreground selectively so that they do not confuse the main subject.

Depth of field is also affected by the choice of focal length (see above). Do not use the term "depth of focus" as that refers to the corresponding condition inside the camera body.

One of the weaknesses of digital cameras is that lenses often do not stop down to f/16 or smaller. However, you will probably not find this a problem in practice. If you use the aperture priority feature of your camera, you can maintain the depth of field required. For great depth of field, the shutter speed may be so low that you may need to use a tripod.

SLR cameras have a depth of field preview button that stops down the aperture iris so that you can check the depth of field in the viewfinder (or LCD). This is extremely useful, except in poor light when stopping down may make it too dark to assess properly.

Film speed and CCD "speed"

Reference to traditional photography is necessary to understand speed. "Speed" refers to the sensitivity to light of a film or CCD. Faster films are more sensitive and therefore require shorter exposures. However, they tend to have a larger grain size, which may be intrusive and may seriously damage fine details of the image. CCDs are analogous in that at high speed settings the images are "noisy", with intrusive bright-coloured pixels and a coarser appearance.

Speed is measured by ISO (International Standards Organisation) ratings: most film is in the range ISO 100–400, but slower or faster films can be purchased for special purposes. Because of the wide understanding of film speed ratings, digital cameras have analogous

settings. For most purposes, the maximum detail will be delivered by a slower rating.

One of the advantages of digital equipment is that you can change the speed at any time, which cannot be done with film unless you keep detailed notes of each frame and send the film to a specialist laboratory.

As with film, it is advisable, unless absolutely necessary, to avoid high ISO ratings, so that the resultant image is not too noisy. The quality of the CCD will vary from camera to camera but, in general, ISO 100 or 200 is probably optimal for good-quality work. A rating below ISO 100 will result in unacceptably low shutter speeds for many purposes (certainly for the use of long lenses).

Digital image types

Most cameras compress the images using the JPEG (Joint Photographic Experts Group) standard algorithm to give a file with a .jpg suffix that most popular software can process. Better cameras will produce RAW or TIFF images that are much larger and fill up the storage media alarmingly. With better cameras, it is possible to adjust the image both in terms of megapixels and compression. Naturally, the maximum size and minimum compression are best. Small images at maximum compression will give an obvious jagged, "pixellated" effect and will not be capable of much enlargement. JPEG images also suffer from loss of data at each recompression (see Digital image processing and presentation in chapter 7).

As cameras now generally accord with the EXIF (Exchangeable Image File Format) standard, information, including the date and time, aperture, shutter speed, resolution, colour system, etc is recorded within the file. This may be useful for subsequent reference.

Digital image storage

A number of storage types are currently in use. These include Sony Memory Stick (almost unique to Sony cameras), SD (Secure Digital), Smart Media, Compact Flash, xD, and Multimedia. Storage is generally becoming cheaper. You should ensure that you have sufficient storage for your purposes, for example, a 256 Megabyte card will take about 120 images at the standard setting that I use on my current camera. However, the size and JPEG compression ratio of your image will greatly influence file size and, if you use RAW files, the card will only hold a few images.

IBM Microdrives, which are tiny hard disks, will fit many cameras that take Compact Flash cards. At the time of writing, they come in sizes up to 1 gigabyte — enough for nearly 500 of my standard images. However, they are still a little expensive.

It is best to keep your cards away from magnetic fields — such as speakers, televisions etc — as static can ruin the image. As soon as possible, you should install the images onto your hard disk (which is cased in metal), and then onto stable media, such as a CD. For really critical material, make more than one copy.

In-camera metering and the autofocus facility are an unconsidered marvel. Their importance are covered in this section.

Metering and focusing

Light metering

In the "old days" (when I started!), few cameras had any device to assess lighting levels. Photographers used a separate, hand-held light meter, which used a light-sensitive selenium or cadmium sulphide cell to drive a needle that indicated the light value. Sliding scales then allowed one to calculate the correct combinations of shutter speed and aperture for the film speed used. This was gradually superseded by meters fitted to the camera, then by meters actually linked to the camera mechanism. In-camera metering has been the norm for years, and is an unconsidered marvel. In many cases (certainly with single-lens reflex designs), the metering is through the lens, so that the meter reads from the same field of view as the lens.

Despite this apparent sophistication, there are still problems to overcome. Meter readings are based on the assumption that the subject area has an average reflectance of 18%. This is the reflectance value of a standard "grey card", which is used for critical metering purposes. In reality, the subject has areas of highly variable reflectance, while behind it is usually a bright sky. Even the dullest sky is usually bright compared with the subject. The meter reads this brightness as part of the subject, calculates the exposure accordingly, and sets the shutter and/or aperture to suit. Consequently, unless bright sun is behind the camera, the subject is likely to be underexposed, with poor detail in the shadows.

There are two ways of dealing with this. The metering can be set so that it is "weighted", or calculates a setting based largely on a part of the image, usually near the centre ("centre-weighted"). This suffices for many uses, and is often the only metering available on cheaper cameras. More sophisticated cameras will have a range of settings. These normally include "spot" metering, where the meter reading is taken from a very small central area of the image, and "matrix" metering, where the image frame is divided into several areas, which are then averaged. Figures 15 and 16 show how a simple metering problem was overcome.

Alternatively (or as well), most cameras have some sort of exposure lock, which normally sets itself when the shutter is pressed lightly or part-way. This enables one to take a reading from an area with similar tones to the important part of the subject, press the shutter button lightly, then recompose for the desired image before fully depressing the shutter.

If you are really in doubt, and the image is vital, use a "bracketing" technique to take several frames with a range of exposures, one of which is likely to be satisfactory. More sophisticated cameras may have this as a pre-set option.

Figure 15: Wood Street in the City — camera fully automatic

Figure 16: Wood Street — exposure adjusted slightly by using exposure lock

More sophisticated cameras may also have a choice between one-off metering and continuous metering. If the lens is set to continuous metering, there is a constant faint clicking noise as the iris leaves change position. This process inevitably wastes battery power, and conceivably could wear out the iris if the camera is used frequently enough.

If you have a digital camera with a histogram function, you can examine this after taking the image. If the bulk of the shaded area of the histogram lies to the left, the image may be underexposed, while if it lies to the right it is probably overexposed.

Autofocus and manual focus

Few serious cameras are available that have no focusing facility. Such cameras are unsuitable for portraying buildings except as snapshots, and need no further consideration.

Many cheaper cameras will have an autofocus facility only. Typically, the autofocus works by assessing the maximum contrast in the image as the lens travels through the focusing range. This type of focus ("passive autofocus") can be defeated by featureless areas, such as interior walls, and can hunt fruitlessly for a focus point. Optical range-finding techniques or infra-red light beams may also be used ("active autofocus"). This system can be upset by heat sources, black absorbent surfaces and such like.

The results of the autofocusing system must be processed by the microprocessor inside the camera. All this must happen in time for the photographer to capture the subject. With more sophisticated cameras, you can often choose between continuous focusing and one-time focusing. Continuous focusing (in which the lens relentlessly tracks the subject as you change position or the subject moves) is more suitable for action photography. However, continuous focusing uses power, which drains the battery. One-time focusing is more suitable for photography of static subjects, as you can focus the lens very accurately on that part of the subject that must be sharpest, and then reframe the subject as you desire. As well as exposure lock, most cameras incorporate focus lock, so that depressing the shutter lightly fixes both before you finally frame the image.

Most autofocus systems operate from a small area within the field of view, which is normally defined in the viewfinder. This allows the area of accurate focus to be determined. In the case of traditional SLR cameras, the viewfinder usually incorporates additional focusing aids for manual focus. These may include areas that appear to twinkle when the subject is out of focus, a split-wedge range finder, where sharp edges in the subject are in alignment if they are in focus, and LEDs, arrows, or other devices (including beeps or other sounds) to confirm focus.

Only the better digital cameras have manual focusing that is any good, and this depends, to a large part, on the quality of the liquid crystal display (LCD) in the viewfinder or on the back of the camera. If the LCD is too coarse, it is virtually impossible to use it for meaningful focusing. Viewfinders may have a dioptric adjustment so that you can see the LCD at its sharpest.

Manual focus is the most accurate system for critical focusing, especially in close-ups, where it is necessary to show a particular detail with great clarity, for instance wood-borer exit holes, or corrosion on steel reinforcement. The automated system may try to focus on part of the image that is not your prime concern. If this type of subject forms part of your recording activity, you should not hesitate to spend at least £500 (price at the time of writing) on the camera.

Where the focus is particularly critical (figure 17), you will need to pay attention to the depth of field. I referred to this above. Simply put, if you need to record objects in the foreground and in the background with equal clarity, focusing will be particularly delicate. The first essential is to stop the lens down to the minimum aperture. The more sophisticated digital cameras will offer an aperture-priority option. This means that you set the aperture first, giving control over your depth of field, and the camera then selects the shutter speed. In dim light, this may necessitate the use of a tripod. If you do not use a tripod, then your depth of field will be destroyed by camera shake.

SLR cameras have markings on their detachable lenses that indicate the acceptable depth of field at any given aperture. In figure 18, the small marks labelled "16" on the lens show the depth of field at f16 with that lens. This can be very useful in poor light, where the focusing accuracy may not be too clear in the viewfinder or on the LCD. The technique is to set the focus manually at a point that ensures that both foreground and background are within the marked limits.

Figure 18: Wide-angle lens showing focusing marks

A specialist type of lens that is offered for some upmarket SLR cameras is the "perspective control" lens. This is designed specifically for architectural photography. Many of the images that you find in publications such as *Estates Gazette* or architectural books have been taken by professionals who will have used a perspective control facility on the camera, whatever the size of the negative (film is still much used in this field). The perspective control lens can be moved vertically relative to the axis of the lens fitting. This means that the camera may be held perfectly horizontal while photographing a tall building, and the vertical lines are still rendered as vertical in the photographic image (but if you examine the images critically, you may find other distortions have appeared).

More complex professional cameras have a much wider range of "movements" that are not relevant here. However, there are other ways of rendering your verticals as rectilinear in the finished image, as discussed in chapter 8.

Figure 17: Close subjects require precise focus

More useful theory, including shutter speed, colour temperature and using flash, are covered in this chapter.

Colour temperature and colour cast

The light by which we live our daily lives is not always the same colour. The obvious red colour of the beautiful evening sunset is clear enough but, in fact, daylight varies in colour throughout the day. The colour of the light is quantified as "colour temperature".

The scale used is degrees Kelvin. At noon, the colour temperature of daylight is typically between 5200°K and 5400°K. Flashguns produce light output at about this temperature in order to mimic natural daylight. This is why the colours in flash pictures usually look fairly natural.

However, if you take pictures of the same subject in the early morning, midday and in the evening, the subject will appear to be different colours at these different times. If the colour in the image appears to be all wrong, then it is loosely referred to as a "colour cast".

The same term is used to apply to situations where the lighting is compromised in other ways. For instance, if the subject-matter is lit only by daylight from a bright blue sky, typically when the subject is positioned in exterior shadows, it will appear in the resultant image to be bluish (figure 19). Fluorescent lighting often gives a greenish cast because it does not produce a continuous spectrum of light.

These factors were the bane of the film photographer. In digital cameras, matters are both more complex and simpler simultaneously. They are more complex because there are additional adjustments to be made. However, they are simplified because the adjustment can be made at the time of the photograph or at the computer keyboard later.

Most of the better cameras have menu items for adjusting the colour temperature, often to a variety of common scenarios, including tungsten lighting and two types of fluorescent lighting.

Other factors

Figure 19: You never know what you might find down a drain

Figure 20:
Death-watch
beetle holes

Figure 21:
Leadenhall
market

Figure 22: Camera shake is clearly apparent in the closer view

Using macro

A macro image is one that produces an image on the CCD that is the same size as the subject, or larger. In practice, the term "macro" is used more loosely to refer to extreme close-ups. Not all cameras will offer a macro facility. The lens of the camera is optimised for subjects that are, typically, a metre or more away. The lens elements must be slightly rearranged to produce optimal performance at distances measured in centimetres only. The macro range itself may, on some cameras, be subdivided into "macro" and "ultra macro". Some cameras will focus down to one centimetre, although this has its limitations because of lighting problems.

Close-up and macro photography are ideal for details that have to be seen closely to be appreciated. Typically, it may be used for such items as wood-borer exit holes (figure 20), fine detail in ornamental finishes, such as granite, or details of wear and tear, such as rust flakes on steel-framed windows or fine cracks in concrete. However, as the lens is so close to the subject, it may not be possible to use the camera flash where required, and a separate flash may be necessary. Professionals who regularly take such images may use a ring

flash, which fits around the end of the camera lens to produce shadowless illumination. However, in exterior photography, the intensity of daylight may be sufficient to permit a satisfactory exposure.

Shutter speed

Most of your subject material will be entirely static, so that it will not be necessary to select a particularly high shutter speed in order to "freeze" motion. I have, however, occasionally used a very high speed, where light permits, in order to illustrate falling drops of water from leaks. Indoors or in very dim light, of course, such freezing can be achieved by use of a flashgun.

Shutter speed must not be entirely neglected. An old rule of thumb is that you should use a shutter speed with a number no less than the focal length (35mm equivalent) to ensure acceptable sharpness without camera shake. In practice, this means that the 28mm wide-angle lens requires a 1 / 30 second minimum, while a 200mm lens requires 1 / 250 second. If the light is not sufficiently good for the high shutter speed required (figures 21 and 22), you should use a tripod.

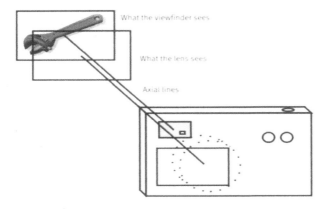

Figure 23: Parallax arises where the viewfinder is separate from the lens

Most cameras now offer a warning light or other indication in the viewfinder if the microprocessor suggests that the shutter speed is too low.

In exceptionally dim light, particularly in interiors, it may be necessary to use the tripod in any case, in order to permit a long exposure. Long exposures may be preferable to the use of flash, particularly in very large spaces (see the discussion on flash technique on page 18). The disadvantage, naturally, is that any objects (usually people) moving through the subject area will produce blurs that detract from the image.

Viewfinders

Cheaper cameras (and even some expensive ones) have a viewfinder that has its own optical system. This is referred to as "direct vision". Because it is not on the same axis as the lens, it sees a slightly different view from that seen by the CCD behind the lens. This is not important until you are relatively close to the subject, when you may find that the difference is sufficient to cut off part of what you saw through the viewfinder. This is termed "parallax error" (figures 23, 24 and 25). Parallax error occurs only with direct vision viewfinders.

Although the lens is very close to the separate viewfinder, it does not have precisely the same view (figure 23). This may be significant for closer subjects and often causes errors, such as cutting off the top of the head in portrait subjects and / or introducing some extraneous item, such as a power socket into the field of view. For relatively close views, therefore, you should not use an optical viewfinder unless it operates through the lens. Even relatively cheap digital cameras have an LCD on the rear, which can be used to frame the image accurately.

SLR cameras (both film and digital) have a hinged mirror and a complex prism ("pentaprism"), which together provide continuous optical viewing of the image until the shutter button is pressed. At that moment, the hinged mirror flies up (usually with an audible *clunk*) to reveal the focal plane shutter or CCD to the lens, the lens iris stops down, and the shutter operates.

Using flash

For about 30 years, a flash tube has been a standard fitting on cameras. An electronic circuit stores charge in a capacitor at very high voltage. When the image is taken, this is then discharged, through a small tube containing xenon gas. This charge ionises the gas and allows it to conduct electricity from a larger capacitor, which has already been charged. This produces a brilliant, but very short-lived, burst of light.

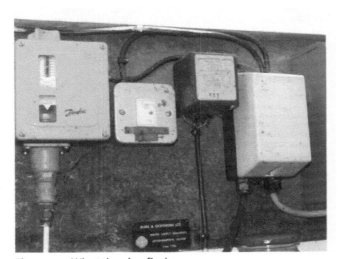

Figure 24: What the viewfinder sees

Figure 25: What the lens sees

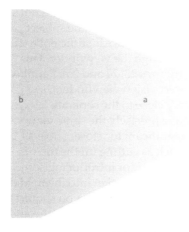

Figure 26: Flash falloff — 'a' is half the distance of 'b' but receives four times as much light

Figure 28: Flash interior

Figure 27: Flash falloff in service duct

A flash gun is essentially a point source of light. Point sources of light are subject to the "inverse square law". This means that the intensity of the light falling on the subject varies according to the square of the distance from the flash to the subject. Or, to put it more simply, an object two metres away receives only a quarter of the illumination received by an object one metre away (figure 26). This is why in flash pictures the foreground is often horribly overexposed, while the background may be nearly black (figure 27). Typically, at ISO 200, the flashgun built into a relatively simple compact camera will have a range of about six metres before illumination falls below useful levels.

It follows that, unless your subject-matter is all at approximately the same distance, there will be exposure problems in the resultant image. Therefore, it may be better to photograph an interior of any depth with natural lighting. However, this may bring problems of colour temperature according to the type of lighting that may be present within the room.

Walls with good reflectivity may channel the light towards the rear of the area of interest by reflecting it several times (figure 28). Examination of your digital image will show if you have been fortunate.

On simpler cameras, the flash will have a single output. On complex cameras, and on many higher-quality separate flashguns, the output is variable. It may be controlled directly by the metering system of the camera, by way of a sensor inside, or on the front of the camera that turns the power off when exposure is sufficient.

You should study your instructions carefully so that you so fully understand the effective range of the flashgun.

In very simple cameras, the flash may be switched on automatically if the ambient illumination is too poor to allow you to take a shake-free picture. This is why, at night-time sporting events, the seating areas produce a constant twinkling of tiny flashes. These cameras are not good enough for your requirements. However, many users of better cameras, who have not taken the time to understand them, inadvertently leave the flash switched to automatic, and this results in the same problem. Do not use the flash unless you really need it.

Because the flashgun is part of, or fixed directly to, the camera, it produces axial lighting. This results in flat illumination with virtually no shadows to show relief. Where the subject-matter is differentiated largely by colour, this will not be so important. However, if the subject depends upon texture for its visual meaning, then that meaning may be lost. Fine details, such as cracks, may be filled in by the flash and become effectively invisible. Separate lighting is a matter for the professional, although I have had some success in using a million-candlepower torch (only in interiors) to provide side-lighting for such details. Small torches are usually swamped by the flashlight.

Flash can, of course, be used out of doors. There are two main situations in which it can be useful. Remember, however, that without an interior with reflective walls, the power of the flash will be partly wasted.

You may find that the natural daylight is throwing deep shadows onto parts of the subject that you need to show adequately. If you are close enough to the subject, it is possible to use the flashgun to "fill in" the shadows to some extent. There are still shadows in the image, but there will be more detail in them.

Alternatively, where the natural exterior daylight is too dim for adequate exposure to allow you to take shake-free images except by using a tripod, the flash can simply be used to provide adequate light. However, you must make sure that you are within the effective range of the flashgun, otherwise the resultant image will be muddy and underexposed.

Using tripods and monopods

Unless you are able to use really high shutter speeds (1/250 or greater), a tripod will increase your sharpness. Below 1/30 second, it is essential for critical work. Rigidity is essential: a tripod that wobbles in a stiff breeze is of little use. However, the more massive and stiffer tripods are heavy to carry around and if you are not taking pictures every day, you may prefer to avoid the clutter of a full-size tripod. Very small tripods, some with flexible legs, are obtainable and the smallest ones can be stowed in a large pocket or within a briefcase. These small tripods can then be set upon a coping, desktop, or other convenient raised surface. Use the self-timer, as advised below, to avoid camera shake.

Once the camera has been mounted on the tripod it is essential to check that the tripod head is horizontal. This is done by looking through the viewfinder very carefully. It is all too easy to overlook this matter, and unflattering skewed images may result (although digital images can, of course, be rectified and cropped to suit subsequently).

A tripod is essential when producing broad, panoramic images. Successive digital images can be "stitched" together to produce such an image. Some cameras may offer some type of assistance system, such as showing the previous image at a reduced size in the viewfinder to assist your composition. The stitching can then be achieved in the computer. With such procedure, of course, horizontality, and therefore the careful alignment of the tripod, is absolutely vital.

It is necessary to use the self-timer when using a tripod to avoid jerking as you press the shutter. Typically, this runs for 2 or 10 seconds. It is not usually for self-portraits, unless you want to get into the image to point at a detail.

Monopods are similar to tripods, but with only one leg. They can be used to steady the camera by pressing down on them, and quite long exposures can be successful. Their other virtue is their small size.

Having an extendable rod to hand can also be useful for other surveying purposes, such as tapping the ceiling in order to gauge its construction.

Maintenance

● Keep the camera safe in accordance with the manufacturer's instructions.
● Keep it away from dust and moisture (fungi can grow inside lenses!), sunlight and high temperatures, and strong electromagnetic fields.
● Keep it clean with a selection of blower brushes (little brushes attached to a puffer), proprietary wipes, and very old, much-washed cotton handkerchiefs.

Full understanding of your equipment is essential, especially with highly specified cameras and this chapter gives some helpful tips for the surveyor operating on site in the real world.

The camera as a notebook

As each individual digital frame is costing next to nothing, you can afford to fire the camera off in all directions with impunity. Although the camera cannot be a substitute for carefully made notes, it does provide a rapid *aide-mémoire* of details that you might not otherwise note, and is a ready reference for the layout of such items as fire extinguishers (figure 29), signs, car park markings, and the like.

You can also use the camera to photograph documents or plans that may come unexpectedly to hand, but in such case beware of two things:
● on-axis flash will often produce a bright "hot spot" on the paper
● any creases will be much more obvious in the photograph.

Figure 29: Fire precautions quickly noted

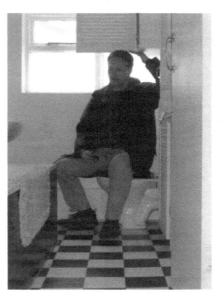

Figure 30: Getting yourself in the image

Technique in practice

Figure 31:
Reflective lift
interior

Figure 32: Image obtained by lowering the camera into a raised floor void

Putting yourself in shot

The self-timer is essential when using the tripod, but can also be used to include yourself in the image. You can then demonstrate something to the camera by pointing, holding a spirit level, etc. In figure 30, the author demonstrates a potentially lethal wc/boiler cupboard arrangement. The camera was parked on top of a convenient radiator. This is more dramatic than reporting that the cupboard is only "x" centimetres above the pan.

Beware that mirrors, or other reflective surfaces opposite the camera position, will reveal the photographer. If the flash is used, a mirror or bright finish will reflect a bright hotspot (figure 31), which will detract from the image and make it seem amateurish.

Use a filter

If your camera can take a filter (all SLR lenses will), use a UV (ultraviolet, often termed "skylight") filter to protect the front lens element against scratches and dust. Grease or other smears on the filter can be scrubbed away with more vigour than would be safe for a lens. The filters cost only a few pounds and can be discarded if damaged.

When fitting the filter, ensure that no dust is trapped behind it, as this can produce annoying blurred spots on the image.

Perspective

For reasons that are unclear, the combination of human eye and brain differentiates between reality and the printed or screen image. When you look upwards at a tall building, perspective makes the vertical

elements of the building appear to converge towards a vanishing point. As you know that these elements are vertical, you think nothing of it. However, when you view the finished photograph taken from the same viewpoint without any adjustment, you may find the converging verticals disturbing. This is why, since early in the development of photography, professional photographers have used cameras (for architectural subjects) that have a range of movements that enable them to render the verticals as vertical in the resultant image. This has become a convention. However, it results in the top section of the building being progressively stretched, introducing a different type of distortion.

There are two ways in which you can deal with converging verticals that may be unacceptable. One is to take a picture from a distance using a long-focus lens. Alternatively, if there is no suitable distant viewpoint, which is likely to be the case in most urban situations, you can use a wide-angle lens or zoom setting and frame your image so that the building is in the upper part of the image only. This will probably require turning the camera sideways. You can then crop the resultant image (obviously easier with digital equipment) to remove the unwanted foreground.

Confined voids and fine details

When confronted with small hatches or other narrow apertures with voids beyond (figure 32), one can insinuate the camera (with great care — and be sure to wind the strap round your wrist) and fire off flashes blindly. The results can be immediately inspected and may be adequate or may suggest further attempts. In such situations, focusing in the dark void is often a problem.

Figure 33: Fine shrinkage cracking

Figure 34: Water collects in hollows and shows them up

Some cameras have hinge-out and / or rotatable viewfinders and these can be used to peer round corners, or over walls, at arm's length. Again, keep a firm grip.

This type of viewfinder can be used to take pictures surreptitiously of, for instance, unlawful activity by tenants. Most people assume that if the camera is not at your eye, it is not taking a picture, so try using it at hip height (and turn off the beeps).

The eye can pick out and accurately assesses cracks (figure 33) or other fine details that are barely resolved by the combination of the camera lens and CCD. As such, it is necessary to make sure that you have recorded the subject properly by checking the image. If necessary, move closer. I have read dozens of schedules of condition that refer to cracks that are completely invisible in the attached photographs.

Finally, when illustrating details, put a measuring rule or tape, or other familiar object, next to the subject. The scale may not otherwise be apparent to a third party.

Rain and shine

Use wet weather to your advantage
While you may not immediately think of taking pictures in wet weather, in fact, it can be quite beneficial from the point of photographic records. For example, puddles (figure 34) record the position of uneven paving far more graphically than any measurements, while deficiencies in rainwater disposal arrangements can be readily recorded. Also, water staining of external building features reveals the behaviour of water relative to copings, damp-proof courses, and such like. Finally, blocked rainwater gulleys or surface water gullies are immediately apparent in photographs.

Lighting
The advice given to absolute beginners is perfectly sound: keep the sun behind you. If you have little or no choice of position, use a lens hood, foliage, shop signs or even a hand to keep the sun off the lens. If you are using a through-the-lens viewfinder or the LCD, you can check the effect. Make sure that you are metering for the subject and not for the background.

Strong cross-lighting will bring out textures, such as hammered concrete and mellow brickwork, and will also show cracks and bulges.

Composition techniques

Cut out the clutter
Particularly when taking quick exterior shots of buildings, it is all too easy to lose track of your objective and find, when you examine the image later, that the foreground is full of cars, people, railings, street signs and other clutter (example on page 30). While it may be impossible to remove all of these, you should seek a viewpoint that enables you to avoid such a foreground.

An elevated viewing position is often helpful, if possible. Also, the use of a monopod and the self-timer with the camera held a metre or more above your head should enable you to remove many extraneous objects. If the image is unsquare as a result of this technique, it can be rectified on the computer if you have used a digital camera, or if you scan a film image into the computer.

Background clutter can also detract from the image, and it may be possible to deal with this by throwing it deliberately out of focus. Otherwise, you could resort to changing your viewpoint or resigning yourself to making the distracting background fuzzy in later computer editing.

Figure 35: Looking almost vertically upwards at a well-known City building

The artistic touch

Even if you have taken a picture that is adequate for its main purpose, you may be able to improve it slightly by careful composition if you wish to give a good impression of a property. A fountain or reflective pool, a sculpture, or strategically placed soft landscaping can dress up a picture that is otherwise purely functional. In such cases, be sure to check your depth of field.

On the other hand, in the case of dilapidations or breach of covenants, you may wish to ensure that the problems are adequately illustrated. In this instance, depth of field can be exploited to show dumped material in the foreground, with the building still sharp and readily identifiable in the background.

Sometimes, the use of a deliberately unusual viewpoint can be used to emphasise a particular feature of the building that a conventional view does not render suitably striking, or which the casual viewer of the building would normally take for granted (figure 35).

Finally, the use of carefully selected filters may improve some images. For instance, a polarising filter can make a pale blue sky more intense. Any number of photographic handbooks will outline further composition techniques.

This section outlines what can be done with your digital images.

Image processing and presentation

The ability to check your image immediately after taking it means that you should have a reasonable set of pictures when you return to your desk. If you are already practised with the digital image, you should be able to use most of your images immediately without any editing. However, you may wish to give some of them further treatment.

Warnings to heed:

● JPEG images are "lossy" and each time the file is saved there is a loss of data. When editing, work on a copy of the original, or save it as a TIFF (.tif) or other format before finally recompressing the image.

● TIFF files are enormous and can soon fill up a hard disk. However, RAW files are even bigger.

Most digital cameras are sold "bundled" with one or more sets of software that enable the images to be uploaded speedily to the computer and thereafter processed with varying degrees of sophistication. However, you may prefer to purchase suitable image-manipulation software separately.

Once you have opened your image in the appropriate software, you will be able to carry out the following functions in most software suites:

● Adjust the colour balance.

● Adjust the exposure (within limits — if you have really made a mistake then there may be no solution).

● Adjust the contrast (again, within limits).

● Remove blemishes from the image (obviously, there must be some limitations — for instance if a person's head really obstructs a significant part of the image, the shot will need to be retaken).

● Reduce graininess in scanned negatives.

● Resize and crop (remove parts of) the image.

● Change the perspective of the image (within limits, otherwise the result may look unnatural).

● Apply sharpening to make the image look crisper (you cannot add what is not there, but it is possible to make the image look more acceptable).

Most of the above tasks can be carried out by a few simple steps. However, it is possible to carry out much more sophisticated editing, such as cutting out the background completely if it is too cluttered, adding graphics, such as arrows, circles or other indicators and so on. You could, for instance, impose your company logo in a corner of the image.

A whole menu of specialist tools is available, usually from a draggable "palette". This allows you to select individual parts of the image manually, by lighting level or by other techniques, so that these areas can be manipulated without affecting the rest of the image. Most software comes with tutorials or online help. There are also numerous textbooks available in any sizeable bookshop.

After taking the picture

Do not forget to save your edited images, as well as your originals. Use a folder system that will enable you to retrieve them later if necessary, and make sure that they are backed up securely.

Never edit the original but save it under a different name, and then edit the second image.

Displaying images

In a report
Space does not permit full treatment of this subject but a few points deserve mention.

As far as word processing is concerned, Microsoft Word is the standard for most offices. Pictures can be inserted readily enough, but positioning them next to the desired part of the text, and wrapping the text around them, may not be so easy, especially for larger images. Particularly when time is of the essence, it may be simpler to create a new section that contains all the figures, which can be referred to within the body of the text.

Microsoft Publisher, which is also part of the Microsoft Office suite, uses "frames" to contain images and text. These frames can be precisely sized and positioned on the page, and captions can be placed anywhere in their own frames and locked to the image frames.

If your document is to be published, for example as a brochure, remember to make sure that all the images will print at 300dpi (dots per inch), or the typesetters will probably ask for new images.

Remember to ensure that the image size is suitable for the documents that you are producing. Carry out all the editing of the image files before setting up the document, save them under another name, and then reduce them in size. If you intend to work with files that are oversized for the document requirement, your document file will be huge and will be slow to save and reload.

It is advantageous if you can print to, or save your file as, an Adobe PDF file. This format can be read by virtually any computer (the Adobe reader can be freely downloaded from Adobe and many other websites, and often comes as standard with a computer as part of the "bundle"). This means you can easily send your report by e-mail, provided the connection has adequate bandwidth. This has the further advantage that a PDF file is usually smaller than the original file. However, you may need to adjust the parameters of your PDF file settings to get the best possible results.

On the web
To do this yourself, you will need administrative rights. If you need to load your image on the internet as part of a web page, first remember that viewers may not have broadband access, so you should make the

Figure 36: Flatbed scanner

image file size as small as practicable. GIF (.gif) files, which use only 256 colours, may be preferable for small viewing sizes. If you are concerned that a reasonably large version of your image should be available, it is probably better to insert a thumbnail on the relevant main page with a hypertext link to the full-size image. You will need to study the online help or software manual in any given case.

Scanned images

Using a flatbed scanner or film scanner...
The flatbed scanner (figure 36), which is probably mostly used for documents, can be used for photographic prints so that you can load the images onto your computer. However, you will need to make appropriate adjustments on the computer so that the print is scanned at a suitable resolution and/or magnification. Therefore, it is essential that you understand the combination of the scanner and its software beforehand in order to avoid disappointment. In addition, this process can be very time-consuming since only one image can be dealt with at a time and so it is best avoided if at all possible.

Film scanners are also an option. The film scanner deals with negative strips, which are automatically converted to positive images during the scanning process. These can then be processed as if they have come directly from a digital camera. This is, however, a specialist unit outside the scope of normal office life.

... and putting them on a photo CD, picture CD
Today, many film processors offer the option of supplying a CD of scanned images as part of the film-processing package. Although the quality of the CD images may not be great compared with the result from a digital camera or from a film scanner that has been carefully operated, it may be possible to improve matters at your own computer.

Album software

Your digital camera software package may include an album-type database, such as Adobe Photoshop Album or Paint Shop Album. While these are easy-to-use products that have been designed for the mass market, they are, nevertheless, very useful and will allow retrieval of images by date, subject or other criterion, together with captioning. Such software is also cheap, and while it may not offer the sophistication of a purpose-made photographic database, it is usually well debugged and works "out of the box".

Album software normally works by forming links to your images so it knows where each individual image is located on your computer. It also recognises the file format. All this information is held in a database. Normally, the images will be presented as thumbnails for browsing but if you want to examine an image closely, the software opens the full-sized image.

Once you have loaded your images into the album database, you can then attach labels or tags that might give such information as the property address, date, or type of building, just as with any ordinary database. Data entry is very easy. The album software usually allows the display of the EXIF data, including the date, which is part of the file. It may be possible to open more than one catalogue so that more than one person can use the same package, subject to licence rights.

You can leave the original folder arrangements untouched so that they can be archived in the order in which they were taken, or you can use the album software to acquire the images directly from the camera. It is advisable to avoid possible conflict between the album software, the camera software, and any image manipulation software as not all packages may be compatible. Finally, if your camera comes with a full set of software, it is probably better to use this unless you are very experienced.

This chapter highlights a few real-life examples to show what can be done using common software packages.

Converging verticals

This building (figure 37) presents an unsightly case of converging verticals if photographed without thought. Standing back as far as possible, I used the 28mm lens and framed the subject in the top half of the field, taking care to get the verticals as parallel to the edge of the frame as possible (figure 38). I then cropped the lower section of uninteresting pavement to give a simple, rectilinear image (figure 39).

Figure 37: Original image with converging verticals

Figure 38: Converging verticals removed by holding the camera straight

Simple example projects

Figure 39: Foreground cropped away to give final image

Figure 40: Image shot without thinking clearly

Figure 41: The same building re-photographed after a little thought

Foreground clutter

As with many buildings in town centres, this new development (figure 40) can only be seen with a proper perspective from unsatisfactory viewpoints. Moving forward and standing as close to the edge of the pavement as possible, I waited until all pedestrians were reasonably far away and the traffic lights had stopped most of the traffic whirling around the one-way system (figure 41).

Retouching to remove an intrusion

Using the clone tool (or equivalent) in your software package, it is easy to carry out simple retouching. More sophisticated retouching is best left to the expert.

This picture of 55 Broadway (figure 42) is spoilt by the intrusive flagpole protruding from a building on the right. First, I used the eraser tool to remove most of the flagpole and halyards, as the sky is white. This inevitably meant removing part of the right-hand distant flagpole on 55 Broadway. Figure 43 shows part of the image erased.

Second, I used the clone tool to copy parts of the building on the right and to paste them over the parts of the image occupied by the flagpole and halyards. I have deliberately left this part of the task unfinished so that you can see what I have done (figure 45).

Third, I used a masking tool to copy the top part of the left-hand flagpole on 55 Broadway and paste this onto the "stump" of the right-hand flagpole. Using a mask allowed me to resize the copied section of the image. Figure 44 shows the copied part of flagpole "on its way" to the new location.

This simple piece of editing took about 15 minutes. It is not perfect but would be adequate for insertion in

Figure 42:
55 Broadway

Figure 43: Editing commenced

a report at a size of 100mm or thereabouts. If you are able to spend, say, an hour on the project (which is well worthwhile for a critical picture), then you can produce a near-perfect result.

Adjusting perspective

As with many city-centre buildings, the "Gherkin" is difficult to photograph from ground level as there are few good viewpoints. I shot the building from outside Lloyd's. Even with a 28mm lens, it was impossible to avoid obvious converging verticals (figure 46).

At the computer, first I cropped the right side of the image to centralise the tower. Then, I used a perspective control tool to "pull out" the two top corners of the image so that verticals are parallel to the frame edge. However, this has the effect of making the Gherkin look fatter than it is (figure 47). Finally, I resampled the picture to make the whole image taller, restoring the apparent height and slenderness of the tower (figure 48).

Figure 46: Swiss Re — original image

Figure 47: Swiss Re — rectified image

Figure 48: Swiss Re — reproportioned

Figure 44: Work in progress

Figure 45: The final result

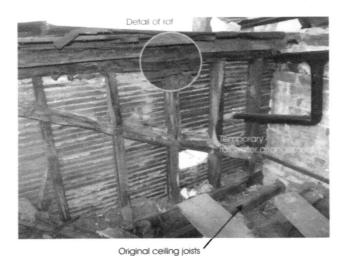

Figure 49:
Annotated
image

Annotating an image

It is easy to make any image more informative by adding text comments, arrows and other features.

Figure 49 shows the inside of the upper part of a shopfront of unusual construction. The timber structure that supports the lath and plaster façade is badly rotted in several places. I imported the original JPEG image into a graphics program. Then, I drew a circle around part of the rotted timber top plate and used a lens tool to magnify a section of the image inside that circle. I altered the line colour of the circle to yellow to make the lens stand out.

I have used two labels to indicate other features; you may prefer to place captions inside the image or outside it; I have done both. It is important that the thickness of the callout lines or arrows is adequate in your final product; it is easy to make the mistake of drawing a hairline that will be indistinct, or even invisible, in the finished report or other output. When the background is dark, a white or yellow line colour is usually most effective. Your graphics program may permit various types of arrowhead and tail that may be more or less effective with different images.

Practice

Now that you have reached the end of this publication, you should have sufficient guidance to:
● Read more knowledgeably through the magazines and web pages in choosing your camera.
● Be capable of producing reliable images with that camera.
● Be able to start learning, in detail, how to deal with the images and produce stunning reports.

If in doubt, read more magazines and textbooks until you have a fuller grasp. But, above all, practice.

Appendix

Further reading

Books

There are innumerable books on photography but you may not find more than a few titles in any given bookshop. As well as building on the specific techniques outlined in this book, you may wish to improve your photographic technique generally. A warning: older photographic experts are sometimes rigidly opposed to digital for some reason, so be careful in your choice.

Useful starting points are:
● *Keep It Simple Series Guide to Photography* by John Garrett, Dorling Kindersley, 2004.

This is a bit weak on digital photography but packed with useful photographic basics that apply to most situations, in a small, handy format.
● *Encyclopedia of Digital Photography* by Tim Daly, New Burlington Books, 2003.

This is not really an encyclopedia as it is in a narrative style but it contains extensive coverage of image processing using the most important packages. Although becoming rapidly out of date, it is useful nevertheless.

Other books worth consideration include:
● *Understanding Exposure: How to Shoot Great Photographs with a Film or Digital Camera* by Bryan Peterson, Amphoto Books, 2004.
● *Digital SLR Masterclass*, by Andy Rouse, Guild of Master Craftsmen, 2004 (if you want to use an SLR).
● *Digital Photography Pocket Guide* by Derrick Story, O'Reilly UK, 2003.

Software manuals

It is worth remembering that most software comes with online help or tutorials of some sort. Some packages still come with "proper" handbooks, which you can conveniently study away from the computer.

Magazines

Photographic magazines usually include sections on choosing cameras (often with extensive reviews of new models) and improving technique. There is a wide choice, with competition between such titles as *What Camera*, *What Digital Camera* and *Which Digital Camera*, not to mention *Digital Camera Shopper* and *Digital Camera Buyer*. One month's worth of such magazines will supply a huge amount of information.

Magazines dealing with digital photography usually have sections on image manipulation although, unfortunately, many of the projects are gimmicky (on the level of putting heads on other bodies and such like). My advice is to browse carefully at the bookstall.

Index

T - #0120 - 071024 - C44 - 297/210/3 - PB - 9780728204775 - Gloss Lamination